© 2012 by Vital Spark LLC. All rights reserved.

ISBN 978-1-105-59176-1

This book may not be reproduced for commercial use, in any form or by any means, without permission in writing from the publisher. This book may be reproduced for educational/ inservice use only.

If citing this book for academic purposes, please cite as:

Goodwin, Heather (2012). 10 Minute Inservices on Dementia Guide for All Dementias. Vital Spark LLC, United States of America.

Printed in the United States of America

10 Minute Inservices on Dementia

Heather Goodwin MSOT/L

Course Checklist

- ☐ **Neuro Review**
- ☐ **Midbrain/Cortical Sensory Impairment**
- ☐ **Brain Scans**
- ☐ **Principles of Neuron Cell Specialization**
- ☐ **Global Deterioration Scale**
- ☐ **Common Irreversible Dementias**
- ☐ **Rare Irreversible Dementias**
- ☐ **Reversible Dementias**
- ☐ **Dementia Stages**
- ☐ **ADL-Types and Physical Assistance**
- ☐ **ADL- Skilled Cues**
- ☐ **ADL- Tricks of the Trade Part 1**
- ☐ **ADL- Tricks of the Trade Part 2**
- ☐ **Behaviors- A challenge**
- ☐ **Behaviors- Trade Secrets**
- ☐ **Physical Goals and Techniques**
- ☐ **Sensory Deprivation and Stimulation**
- ☐ **Activities**
- ☐ **Information Processing Model**
- ☐ **Allen Battery**
- ☐ **Brief Cognitive Rating Scale**
- ☐ **Staff Education**
- ☐ **Mental Models**

Neuro Review
May 2010

Vital Spark

Dementia may affect any area of the brain.

Cortex/Cerebrum: Processing and Management
- **Frontal Lobe:** impulse, personality, divergent thinking, judgement, organization, volition and emotional control
- **Temporal Lobe:** language expression and reception, symbolism, logic, time, navigation
- **Parietal Lobe:** sensory and motor input and motor output
- **Occipital Lobe:** vision and visual processing
- **Corpus Callosum:** bilateral coordination, symbolism, language, right/left brain connection

Cerebellum: Motor Praxis

Mid Brain: Autonomic/ Primitive
- **Hypothalamus:** regulation of hunger, sleep, emotions, temperature, osmotic pressure and sex drive
- **Thalamus:** gateway for 6 of the sensory nerves and reroutes information to cortical area
- **Basal Ganglia:** primitive midbrain, fight or flight and olfactory nerve
- **Hippocampus:** short term memory and working memory

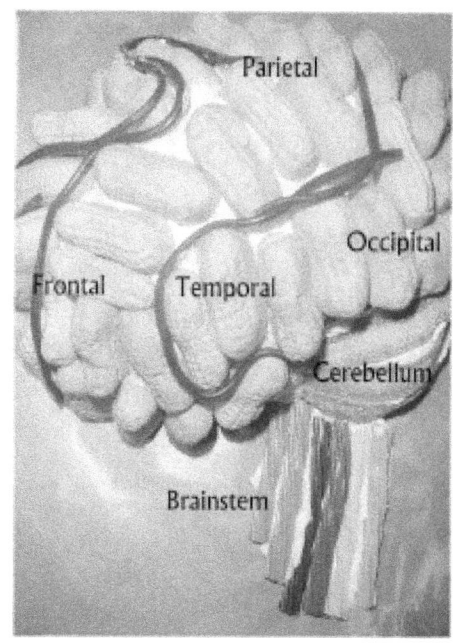

On the evaluation in the cognition section, you would note an increase, decrease or difficulty in any of the listed functional areas

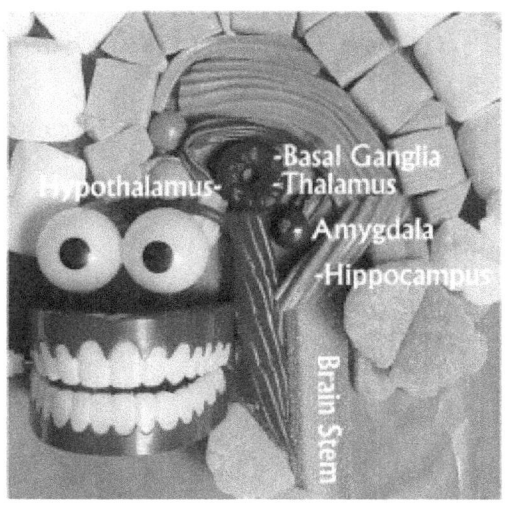

Issue # 1

Sensory Processing

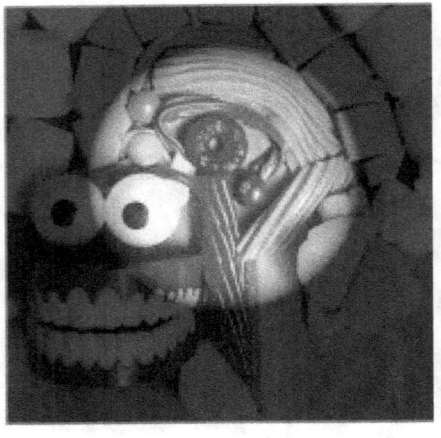 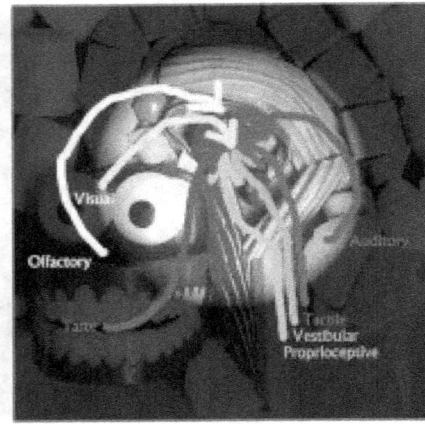 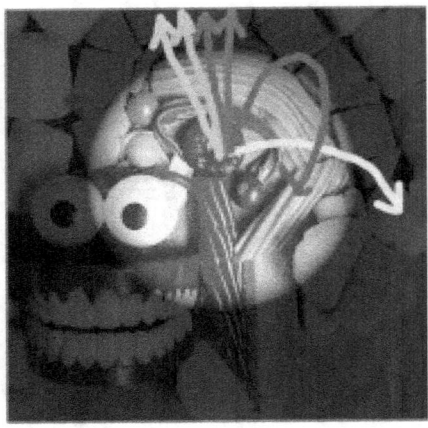

1. Basolimbic Group   2. Receives input from sensory nerves   3. Then sends information to cortex

MidBrain/Cortical Sensory Impairment

It's not a visual problem and it's not a hearing problem, it's a brain problem!

Give 30 seconds to a minute to process the cue

Repeat cue

If quality of cue is not efficient enough to be processed within working memory:

Change the cue

Use multi-sensory cues

Brain Scans

Get something tangible on record!

CT Scan

PT Scans

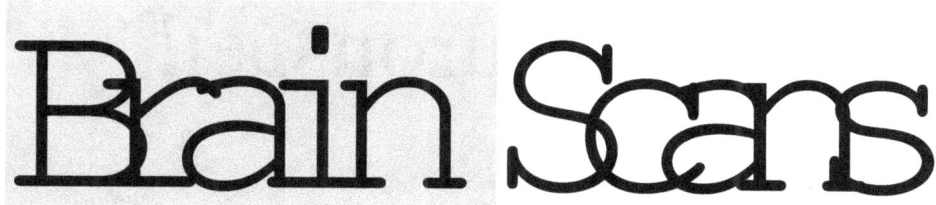

Normal CT Scan   Alz CT Scan

Pictures from http://www.med.harvard.edu/AANLIB/home.html

Normal Brain Activity   Alz Brain Activity

MRI

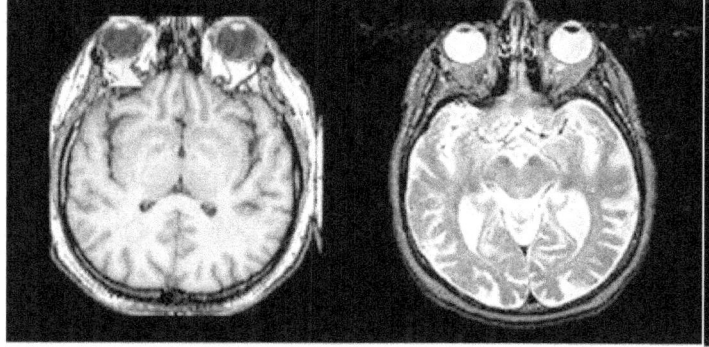

Normal MRI   Alz MRI

Pictures from http://www.med.harvard.edu/AANLIB/home.html

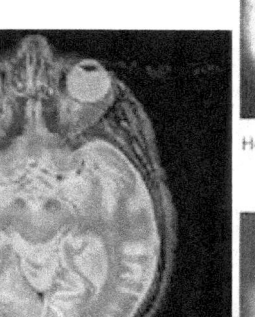
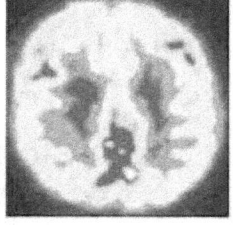
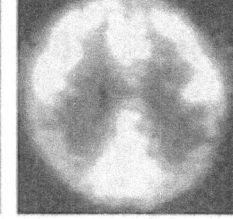
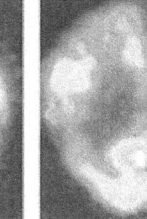

Healthy control   DAT 60 year old male   Severe DAT 64 year old female

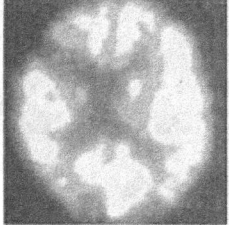
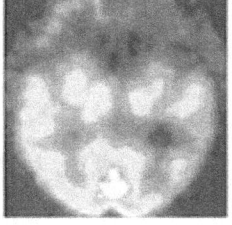
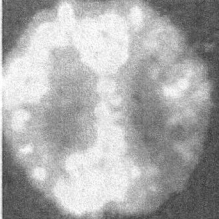

MID (define) 50 year old male   Pick's disease 69 year old female   DAT aphasia 59 year old female

You have the power of referral. Scans may be necessary to determine the extent of vascular damage, rule out rare conditions such as Huntington's and Frontal Temporal Dementias and determine if balance and memory issues are related to CSF build-up. Ask for one if you feel that the information would be helpful in determining eligibility for special therapies and/ or treatments.

The Principles of Neuron Cell Specialization

Once neurons die, they don't grow back
Reginald has forgotten how to put his pants on safely. i.e. while seated.

Neurons create complex networks that connect to each other over several different pathways
Reginald uses an existing pathway to put on his pants. How did he put his pants when he was younger and more nimble? The nurse observes the client putting his pants on while standing.

The more a neural pathway is used, the stronger it becomes
Reginald's nurse tells him to sit down while putting on his pants. He responds by saying, "I don't like everyone telling me what to do! I did just fine when I was home alone. I want to go home!" Would Reginald do better at home? Yes, because routine is plasticity and repetition becomes a substitute for Reginalds new problem solving skills.

Several neurons can cross over the same path
The nurse suggests to Reginald that he puts his socks on first because the floors are cold. While he is seated, the nurse suggests that he put his pants on.

Children have a higher neuroplasticity than adults
A therapist suggests at an early intervention meeting that bulking up the services before a child with severe cerebral palsy. The belief is that the child will make her most rapid gains before she is 5 years old.

The Global Deterioration Scale

(Reisberg, Ferris & Crook, 1982)

1 No Cognitive Decline - *Normal*
2 Very Mild Cognitive Decline - *Age Associated Memory Loss*
3 Mild Cognitive Decline - *Mild Cognitive Impairment*
4 Moderate Cognitive Decline - *Early Stage Dementia*
5 Moderately Severe Cognitive Decline - *Mid Stage Dementia*
6 Severe Cognitive Decline - *Late Stage Dementia*
7 Very Severe Cognitive Decline - *Terminal Stage Dementia*

A "Normal" person demonstrates on average a daily functional performance at a 4.

A person with dementia can also demonstrate a daily functional performance at a 4.

How can you tell who has dementia? By giving a cognitive test that you can make the appropriate inferences from (validity) even if different people give the test (reliability).

People's cognition can wax and wane throughout the day. However, when you present the person with a change or a challenge, you are forcing cognitive performance. Those that can rise to the occasion and get higher scores do not have an impairment. Those that fail or refuse to perform when challenged are a cause for concern to the cognitive evaluator.

How do some of your clients with dementia come out of the woodwork to present symptoms?
- Driving
- Emergency Response
- Recovery After Surgery
- Dehydration
- Over/ Under Medication

 Common

ALZHEIMER's: Alois Alzheimer, 1906

cause: amyeloid plaques, neurofibrillary tangles
types: familial (10%) sporadic (90%)
diagnostic: ApoE-e4, rule out other diagnoses, MRI, MMSE, APP/PS1
incidence: 60% dementia cases. Early onset/ ages 30-65 Late Onset/ ages 65+
prevalence: risk doubles every ten years after age 65. (Starting at 10%) 50/50 inheritance for familial. Higher risk: diabetic, women who gave birth to 5+ children.
Symptoms: Chronic and progressive loss: motor praxis, expressive/receptive aphasia, memory loss. "Retrogenesis"

LEWY BODY DEMENTIA: Friederich H. Lewy, 1912

cause: Lewy bodies in midbrain, brainstem or cortex
types: LBD, *PDD, LBVA, DLBD, CLBD
diagnostic: CT scan, cognitive assessment
incidence: 20% dementia cases. 50% PA develop PDD
prevalence: risk doubles every ten years after age 65. (Starting at 10%)
symptoms: Memory loss, changes in attention and alertness, visual hallucinations, delusions, sensitivity to anti-psych medication, symptoms vary daily

VASCULAR DEMENTIA

cause: vascular damage or multi-infarct
types: Several, including, Punch-Drunk, TBI, MCI associated with CVA etc.
diagnostic: PT, MRI, CT, manual muscle test, cognitive tests
incidence: High risk with multiple TIAs, CVAs, men, football players and boxers
prevalence: 10-20% cases in USA
symptoms: similar to Alzheimer's, although initial cognitive loss is drastic. Lability, incontinence, LTML, STML, dyspraxia, hallucinations, delusions.

Irreversibles — Rare

MULTIPLE SCLEROSIS: Most commonly diagnosed neurological disorder in adults. Disease of CNS (brain, optic nerves and spinal cord). Areas of thick scar tissue (sclera) form on myelin sheaths from lesions/plaques, which interrupt synaptic messages. Characterized by significant functional losses during exacerbations and restoration of some or all function in remissions. Diagnosed by looking at genetic history, autoimmune deficiency and high triglyceride levels. Heat therapies for muscle aches are contraindicated.

AIDS: Acquired Immune Deficiency Syndrome is caused by HIV (Human Immunodeficiency Virus), which weakens the immune system. No longer able to fight diseases, they develop rare pneumonias, cancers and other illnesses. First hit headlines back in the 1970's in LA/NY when incidence was high in homosexuals. Dementia during metastasized cancer stage.

PICK'S DISEASE: A type of Frontal Temporal Dementia. Named after Dr. Arnold Pick in 1892, who noted abnormal protein growths of tau formed brown rounded structures. Often the first symptoms are subtle personality and behavior changes, poor impulse control and language deficits.

HUNTINGTON'S CHOREA: Named after Dr. George Huntington in 1872. Characterized by jerky, almost dance-like movements, children have a 50/50 chance of inheritance from parents who have the disease. The disease routes in DNA as a genetically programmed degeneration of neurons. Symptoms include mood swings, clumsiness, speech slur, STML, difficulty making decisions and concentration. No cure, but you can lessen severity of symptoms through preventative strengthening and conditioning.

CREUTSFELDT-JACOB DISEASE: Variant of Mad Cow Disease, caused by exposure to infected human spinal chord tissue, brain tissue or blood. 90% of cases are caused in this way and are referred to as sporadic. There is a 1 in a million chance of such exposure, which results in less than a one year life expectancy. Acquired is an inherited form of the disease and accounts for 10-15% of the cases. Both are caused by abnormal transformation of the prion protein into abnormal, which is fatal to the nervous system.

WERNICKE-KORSAKOFF SYNDROME: Nicknamed "Alcohol-Related Dementia" because Alcoholics are at high risk for both Wernike's Encephalopathy and Korsakoff Syndrome. Vitamin B1 (thiamine) is used in the glutamate digestion process of Alcohol causing a deficiency. Those with Wernike-Korsakoff benefit from treatments of parenteral Bs to help with absorption. Many symptoms such as amnesia, mental confusion, impaired STM and ataxia can be reversed if treated immediately. Diagnosis can occur after a CBC, MRI, review of liver function or a B deficiency related to other diagnoses such as gastric bypass, AIDS, long term hemodialysis, malnutrition.

Reversibles

Also referred to as Deleriums, are temporary cases of confusion that may resolve on their own or require medical or therapeutic treatments.

Drugs

Emotional (Depression)

Metabolic

Eyes and ears (sensory)

Nutritional

Tumors/ Lesions

Infections

Anemia

ANOXIA
INTOXICATION
AFFECTIVE DISORDERS
UTIs
NUTRITIONAL
METABOLIC
VASCULAR
TUMORS
HEAD INJURY
HORMONES
PARASITES

Comorbidities
Physical Disabilities
Depression
Diabetes
Pain

(Client with these along with irreversible will present with greater cogntive deficits)

Dementia Stages

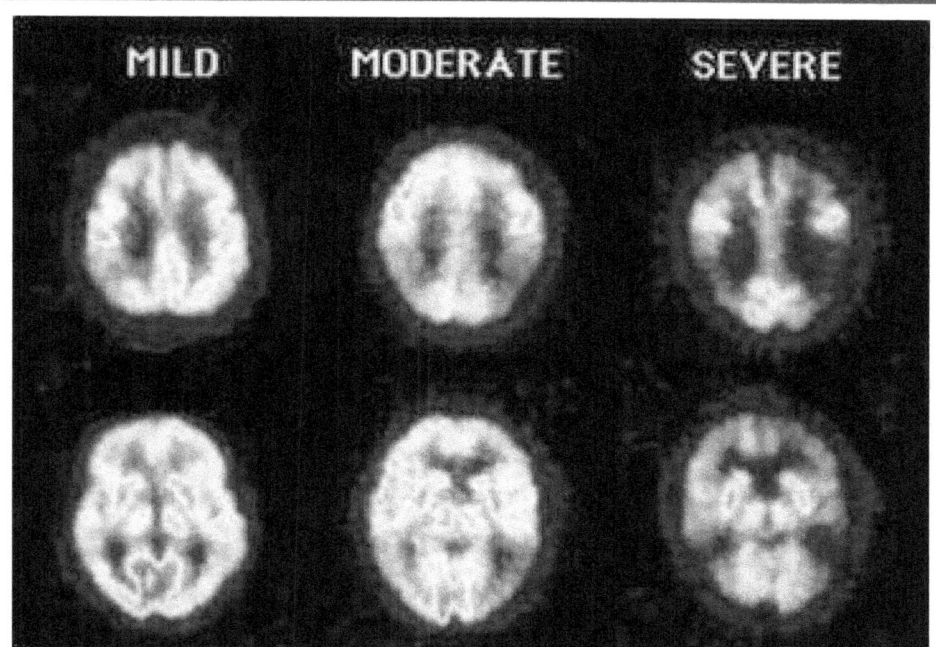

Look at the PT scans to the left:

What kinds of deficits can you see emerge thoughout the stages of the disease process by looking at the following PT scans?

Early
Can live home alone, is independent in self care. Noticeable loss of job skills and organization. Can make plans before actions, be novel, creative and spontaneous. Neuroplasticity of compromised routine may last as long as 30 days.

Late
Needs up to total assistance in all care. Can still groom, feed self, walk and is continent until the very end of this stage. Can participate in repetitive, low level activities (rocking, sorting, twiddling, stirring and activity apron)

Mid
May have lived alone if routine was never compromised by change/ challenge. Neuroplasticity of compromised routine may last as long as 14 days. May need setup and/ or distant supervision in ADLS for safety and sequencing. Most noticeable change in ADLs. Can copy an activity with modeling and learn new concepts within 2 weeks with practice.

Terminal
Transient alertness and rigid muscle tone are observable. Many people in this stage are "bed bound". Careful attention must be made to preserve skin integrity through passive repositioning. Client should have hospice care.

ADLs — Types and Physical Assistance

Activities of Daily Living (ADLS) are self care activities. There are 3 types:

BADLS: Basic
First learned and simple: dressing, eating, bathing

PADLS: Personal
More specific to the person/ culture: health and well-being, hygiene and grooming, medications, vitamins, nutrition and hydration

IADLS: Instrumental
Needs high prefrontal and frontal function: executive management skills, appointments, manners, domestics, emergency preparedness, driving

LETS GET PHYSICAL
When we fill out Evaluations

Levels of Physical Assistance correlate with levels of GDS if a lay person were to give care to a client. Why? Because Healthcare practitioners have specialized cueing techniques and can temper those cues to reduce the need of giving physcial assistance to a client. Your client benefits from these skilled techniques that you use in your treatments.

GDS score	Level of Physical Assistance	MDS	Lay terminology
1	Independent	Independent	Independent
2	Setup	Limited	Leave out materials only
3	Supervision	Limited	Constant/ Periodic Supervision
4	Minimum	Limited	Help with foot care only
5	Moderate	Extensive	Help with lower extermity
6	Maximum	Extensive	Client can groom self
7	Total	Total	Help with all care

* Be aware that Nursing and Therapy use different terminology to quantify physical assistance levels. Write careplans using lay terminology and avoid MDS / FIM descriptions.

 Skilled Cues

Temper your cues to decrease the need for physicial assistance!

Promoting independence is less work for you, it provides the best reflection of the client's dementia level, there is a significant decline in behaviors and you will help improve your client's confidence.

Verbal Cues: Minimum: 3-4 step directions (minimum breakdown)
Moderate: 1-3 steps (moderate breakdown)
Maximum: simple one step (maximum breakdown)
*Most people revert to their primary language in mid-stage

Gestural Cues: 90% of communication comes from your body. Be aware of how your body language communicates with others. Also observe how facial expressions can elicit emotions from others (effective emotions). Think of yourself as being on stage, others are watching how you move and might by trying to elicit cues. Body language is interpreted differently between cultures.

Visual Cues: Signs, lighting, visual contrast, photographs, written instruction

You will know when you encourage independence from your clients, because your brain will be tired at the end of the day as opposed to your muscles.

LAB: Encourage your client to make a peanut butter and jelly sandwich without giving any physical assistance.

ADLs

Tricks of the Trade Part 1

DINING

Problems
- Dysphagia
- Weight gain/loss
- Confussion with multiple items
- Decrease in table manners using fork to cut food

Tricks
- Simple setup
- Food in bowls
- Syrup
- Double portions
- Seat with similar cognition
- Small portions

BATHING

Problems
- Refusal of bath
- Combative
- Not alert
- Refuses to get undressed

Tricks
- Alternative shower/bath
- Blue food coloring
- Rubber ducky
- Repeat approach
- Heat
- Towel robe
- Music
- Spa
- Lotion bath

HYGIENE & GROOMING

Problems
- Mixing up products
- Excessive use of products
- Refusing to shave
- Object permanence
- Perseveration
- Agitated in bathroom
- Urinates in strange places

Tricks
- Line up in sequence
- Grow a beard
- Lighter shades of makeup
- Remove mirrors
- Notes next to mirrors
- Black toilet seat
- Signs

DRESSING

Problems
- Repeats outfits
- Multiple layers
- Sequencing
- Undressing
- Fidgeting
- Refuses to get undressed

Tricks
- Line up in sequence
- Hamper
- Careplan to dress after breakfast
- Change size of clothes
- Have multiples of same outfit

ADLs

Tricks of the Trade Part 2

DRIVING

Taper off with small concessions at a time. Be positive and tell the client what they CAN do as opposed to what they CAN'T.

CANS:
Drive During the day
Drive in town
Drive to specific locations

WHEN YOU ARE WORRIED:
Request a driving test. If they pass, you can request again within 3 months

WHEN THEY CAN'T TAKE NO FOR AN ANSWER: You will need to be creative. Sometimes the doctor will write a prescription to stop the driving

PHYSICAL DYSFUNCTION

Part of the healthcare doctrine of caring for physical disability is the following:

Transfer toward the strong side, Affected side goes in first for dressing, out last.

But you may find that clients with cognitive deficits will have difficulty learning these strategies. You may have to compensate by transferring to the "safe side" or buy larger clothing for your client.

Clients that fall frequently may benefit from wearing hip protectors.

Clients in early and mid-stage dementia may read written instructions on how to safely use a walker. Tape instructions to the walker in a protective sleeve.

Behaviors
A Challenge

ANTECEDENTS: Are the causes of behaviors.
   Psychosocial/Environmental: physical discomfort, over/under stimulation, change in routine/environment/ caregiver (sundowning), task too complicated, inability to communicate effectively.
   Medical: weight loss, fatigue, sleep deprivation, physical changes to the brain, side effects of medication, impaired vision or hearing, misinterpreting the context of a situation, hallucinations.

BEHAVIORS: Some behaviors that are challenging include: combativeness, wandering, stripping, exit seeking, rummaging, sexual inappropriateness, refusal of care, excessive emotions, repetitiveness, urinating in wrong places, uprisings, hoarding, paranoia and yelling.

CONSEQUENCES: May include sensory deprivation, chemical or physical restraint, environment or caregiver change. Physical or emotional harm to client or caregivers.

PREVENTION: Watch for the cues predicting aggression. Make goals to prevent behaviors such as increasing independence, modifying the environment to have natural and consistent lighting, have consistent caregivers and a routine, reduce excess noise, put labels on drawers and rooms (Mitchell, 2002), provide the "just right challenge", don't argue, do acknowledge and validate feelings. Distract client from unpleasant feelings or behaviors with something comforting to do.

INTERVENTION: Some people are defensive or paranoid and may benefit from indirect verbal cues. Indirect verbal cues are a passive way to convey a message to a client while still giving choices and encouragement. One example of an indirect verbal cue would be to say, "After you get out of bed, would you like a cup of tea or coffee". Instead of asking the client if she wants to get out of bed or telling her that she needs to get out of bed, you present the task as a given and lead into a choice of beverages instead.

COGNITION: Cognitive testing is crucial in determining the intention behind behaviors. Make a goal to test and quantify your client's cognitive performance soon after the emergence of challenging behaviors. Can the behavior be modified or do you need to compensate for it by prevention strategies?

Trade Secrets

Behaviors

COMMUNICATION

- *Behavior is a form of communication.*
- *Maintain eye contact at all times.*
- *Talk in a calm and clear manner, using simple terms.*
- *Listen carefully and anticipate what the client is trying to express.*
- *Offer verbal and non-verbal cues.*
- *Watch your body language.*

DON'T

ARGUE
QUESTION
CORRECT
TRY TO CONVINCE
TAKE BEHAVIOR PERSONALLY

DO

ACKNOWLEDGE FEELINGS
ENCOURAGE REMINISCING
REASSURE
REDIRECT

WHEN THE TRUTH HURTS

Try answering a different way

THERAPEUTIC FIB
VAGUE ANSWER
HALF TRUTH
EMOTIONAL TRUTH
VALIDATION
DISTRACTION

CUES PREDICTING AGGRESSION

Verbal Warning: "I'm going to hit you"
Demanding / Threatening / Loud
Clipped or Pressured Speech
Brooding Silence
Short Yes /No Answers
Increased Irrationality
Curses / Uses Coarse Language

Goals and Techniques

Physical

Common Contractures

Neck Flexion
Hand/ Arm Synergy
Leg Adduction

*Daily PROM is crucial, a long slow stretch is just as effective at 10 reps two times a day.

Clear imitation vanilla extract will kill unwanted hand odors.

MAKE IT FUNCTIONAL

Use function to elicit physical movements from clients as opposed to free weights. Clients have more consistent and safe isolated muscle movements if they are related to tasks that are familiar in motor praxis.

INSTEAD OF: shoulder arcs, picking up cones off of the ground, exercise bar, clothespins on a ruler, standing at grabrails.

TRY THIS: stand at sink to groom, hang baby socks to dry on a clothes line, wach windows, sweep floors, pick up spilled cards.

ADAPTIVE EQUIPMENT

WHEELCHAIRS: Have core strengthening goals in conjunction to wheelchair positioning goals. Chose a wheelchair for your client that will cater to a progressive condition: allowing for pedal mobility as well as reclining and assisted mobility. Common abnormal postures include kyphotic, side leaning, hip thrust and extensor tone. When on the fence for recommending a wheelchair, think about the client's mobility both as it relates to physical strengthening as well as independent access of his environment.

WALKERS: For clients that need close supervision and constant safety cues with a new walker, consider training them with a stroller, shopping cart or rollator. When written instructions are not beneficial to cue on using devices safely, consider using praxis instead.

FOOTWEAR: Make goals for client to demonstrate willingness to wear appropriate footwear or to have safe gait with appropriate footwear that will be determined through trials.

Sensory
Deprivation and Stimulation

SENSORY DEPRIVATION: Is a loss of input to the senses that can cause irritability, depression, confusion and a faster progression of the disease. People who have dementia have avolition and do not seek out activities like they once did. We must bring the stimulation to the client, especially if they live in the sterile milieu of a healthcare facility.

SENSORY STIMULATION: Studies show that people with dementia that participated in sensory programs demonstrated increased alertness, decreased behaviors with medical and dental care, increased food intake, decreased behaviors with ADLs and transfers.

NOXIOUS STIMULATION: Is a technique used to help increase alertness in clients that are catatonic, comatose or not alert. These elicit undesirable sensations causing the client to be more alert. Examples of noxious stimulation include: clove oil, lavender, bitter apple, sternal pressure, light touch, deep joint compressions, cold, bright lights and high pitched noises.

The Seven Senses
(In order of complexity)

OLFACTORY: Smell, the simplest sense, related to taste. Important for survival and fight / flight response

VISUAL: Pictures, real objects

TACTILE: Deep touch, light touch, temperature and pain

PROPRIOCEPTIVE: Body sight, comes from the deep joint receptors. Your body can see where your it is in space without using your eyes. Can be triggered by moving or impact.

VESTIBULAR: From the semicircular canals in your ears. Your sense of balance.

AUDITORY: Listening

GUSTATORY: Taste. The most complicated sense, last sense lost is sweet.

Give stimulation, allow 30 - 60 seconds to respond before stimulating next sense in order of complexity. 15 minutes of sensory stimulation can result in up to an hour of alertness.

Activities

It is human nature to be occupied by purposeful activities. Busy people have less falls, less combative behaviors, have a better sleep cycle and have more normalcy with family visits.

1:1 Activities: ADL tasks: Painting nails, grooming, making bed, bringing laundry to laundry cart. Social interaction, short walk on errands, individual attention, nutrition and hydration.

Independent Activities: Anything that a person can stay engaged in independently. The person may need help initiating the activity. Caring for plants and pets, ADLs, sorting, folding, prep work, photo albums and magazines, collection, distribution, rocking chair.

Group Activities: Primary Activities: open ended such as socials, exercise, reminiscence and sensory stimulation. Secondary Activities: can follow rules and take turns such as, BINGO, crafts, dominoes, pet therapy. Swap meet, cooking, Snozelen, Bright Eyes, hydrations, nail care, exercise, walking, setting tables, flea market, Good Will collections, measuring doors, spa treatment, pet therapy.

Resources for Activities:

Alz.org (state chapters) : 101 Activities

AAI-nature.org : Context Box, Bring Me the Ocean

Flaghouse.com : Snozelen and competitors

Aoa.gov : Crafts and music

Allen-cognitive-network.org : Therapist products

Alzheimer's outreach : Activities and suppot groups

Information Processing Model

Cogntive Disability Frame of Reference

Basis for ACL, Assumes that symptoms of mental illness will only subside with result of medication. Therefore therapy focuses more on remedial strategies.

Make recommendations for who you have in front of you. Don't wait for the perfect day to give a cognitive test. Administer and score tests in the standardardized way.

INFORMATION PROCESSING MODEL SCORES

LEVEL 1: Reflexive

LEVEL 2: Movement

LEVEL 3: Repetitive Actions

LEVEL 4: End Product

LEVEL 5: Variations

LEVEL 6: Tangible Thought

IPM Scores are divided into 26 "modes".

*	1.0	2.0	3.0	4.0	5.0
0.2	1.2	2.2	3.2	4.2	5.2
0.4	1.4	2.4	3.4	4.4	5.4
0.6	1.6	2.6	3.6	4.6	5.8
0.8	1.8	2.8	3.8	4.8	*

*.0, .2 and .8 are considered transition phases, .4 and .6 are within a stage and at a stable baseline

There are several tests from the Allen Battery that you could give a client to generate mode scores within the information processing model.

Verify scores through making skilled observations of cognitive performance during daily tasks.

Allen Battery

(Claudia Allen, Tina Blue, Kathy Earhart, Theresa Burns, Noomi Katz)

Allen Cognitive Level screen: Available at S&S Worldwide for $85 USD. This called the leather lacing test,. You teach the client a series of three stiches using leather laces around a punched leather card. The stitches increase in complexity: the running stitch, whip stitch, the cordovan stitch. First offer the ACL, which is the smaller version. If the client has physical or visual difficulties, trade out for the LACL, a larger version. Starting with the LACL only does not change validity. Can take 30 - 60 minutes to administer. Modes 3.0-5.8

Routine Task Inventory- Expanded: Available at the Allen Cognitive Network for free. Recently Noomi Katz has expanded the test with help by Sarah Austin to develop the testing manual. It is a questionnaire and tool for gathering skilled observations to determine scores through a mean score developed by averaging client and caregiver report of skills. Can take up to three half hour sessions to administer. Modes 1.0-5.8, it is suggested to gather parallel scores from the functional independence measure (FIM).

Sensory Stimulation Kits I and II: Currently the kits are unavailable for purchase, but the checklist of skilled observations can be found in the book: <u>Occupational Therapy Treatmment Goals for the Cognitively Disabled</u> (Allen, Earhart and Blue,1992). Modes 0.2-3.2. Administer if the client doesn't take the leather lacing test and cannot write his name.

The Allen Diagnostic Module: Available at S&S Worldwide ssww.com. 27 standardized craft tasks, used to determine modes. Sarah Austin completed her dissertation proving validity and reliability of the placemat task (modes 4.0-4.8) and the Tile Boxes (3.0-4.6).

Cognitive Performance Test: The original test has 6 common daily living tasks that you ask client to perform. In 2008 the "medbox" task was added. http://www.ot-innovations.com has a detailed description. $479 USD at allegromedical.com

Brief Cognitive Rating Scale

5 Axis Categories: concentration, recent memory, past memory, orientation and self care

Rate scores from 1-7, then add scores. Divide by 5 to determine GDS score.

Rough translation of Modes to GDS

Mode	GDS	age
0.8		
1.0		0-1 month
1.2		1-5 months
1.4		4-8 months
1.6		4-10 months
1.8		6-12 months
2.0	7.0	9-17 months
2.2		10-20 months
2.4		
2.6		12-23 months
2.8		
3.0		18-24 months
3.2		
3.4	6.0	
3.6		3 years
3.8		
4.0	5.0	4 years
4.2		5 years
4.4		6 years
4.6	4.0	
4.8		
5.0		7-10 years
5.2	3.0	11-13 years
5.4	2.0	14-16 years
5.6		17 years
5.8	1.0	18-21 years

Staff Education

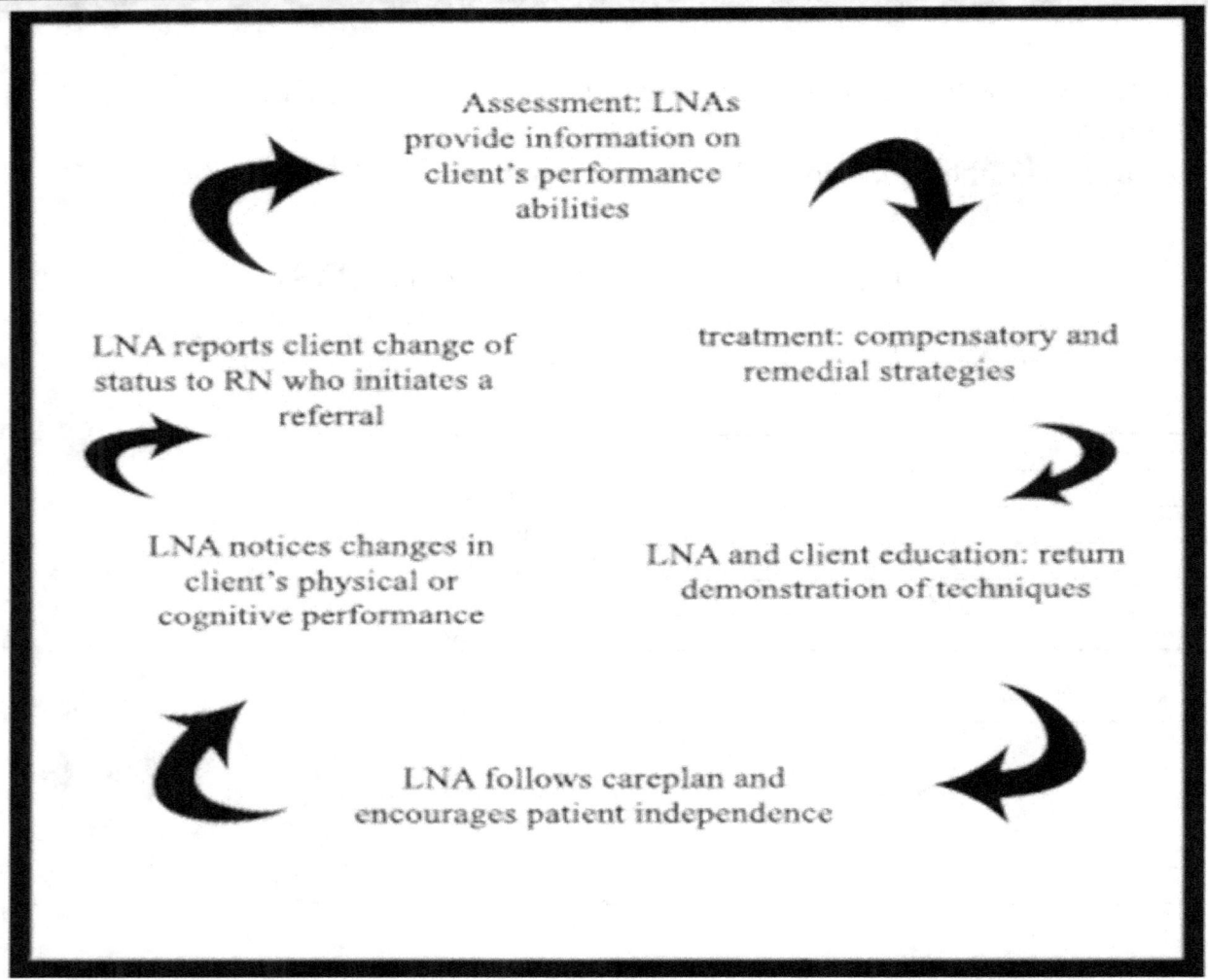

Have weekly contact with staff / caregiver. This will help decrease referrals for repeat issues. Get into their heads by understanding how they feel and learn. Start training of caregivers within last 2 weeks before discharge to ensure return demonstration of techniques. Communicate with all involved parties because everyone is responsible for care.

Mental Models

I N   T H E I R   H E A D S

- *Medical Model*
- *Social Model*
- *Empathy Drain*
- *Golden Rule*
- *Mother Wit*
- *Humor*
- *Hands On Activities*
- *Teachable Moments*
- *Career Ladder*
- *Recognize Excellence*

Golden Rule
Ethic of Reciprocity

Mother Wit
"The client is like my own child"

Medical Model
Every problem is psychological or physiological

Social Model
Everyone is equal

Mental Models

Are the way that we make sense of the world. They are the way that we make sense of the world, are incomplete and contantly evolving. Studies show that professional clinical knowledge doesn't have the primary influence on thought process, sense-making and how we interpret the consequneces of our action (Anderson et al, 2005).

Notes

Notes

Notes

Notes

Notes

www.ingramcontent.com/pod-product-compliance
Lightning Source LLC
Chambersburg PA
CBHW081147170526
45158CB00009BA/2755